OPC UA: The Basics

An OPC UA Overview for those who are not Networking Gurus

JOHN RINALDI

ISBN-13: 978-1482375886
ISBN-10: 1482375885

DEDICATION

To the Automation Engineer, the unsung hero of American Manufacturing.

TABLE OF CONTENTS

ACKNOWLEDGMENTS

This book would not be possible without the dedication, friendship, persistence, support and follow through of the entire staff at Real Time Automation. Specifically, many thanks to Jeff Stiefvater for his advice, support and friendship. And endless gratitude to Drew Baryenbruch for freeing me of daily sales and marketing so that I can take on projects like this.

INTRODUCTION

Why a book on OPC UA?

In 2011 I started reading and studying OPC UA, the successor to the widely popular OPC for MS Windows (what I now refer to as "OPC Classic"). There are two really good books on the subject written by some of the authors of the OPC UA specification (see resources section). I've read both these books.

I have no complaints with those books, but I find them exceptionally pedantic. They can be hard to read and require a lot of patience. Connecting the dots from what's in one section to something in a later section is difficult to do without exceptional diligence.

Even though those books are more than adequate, I felt that there were two good reasons to add my contribution to the subject. First, I felt that it was important to explain UA in a shorter, more direct way. Those books are fine for software developers and people who really want a strong background in UA. But that's certainly not most people. Most just want to figure out what this is all about and don't need 350 pages to do that.

Secondly, I wanted to place UA in context for my particular customers: end users, system integrators and machine builders in industrial automation. Many, many of these people are somewhat insulated from the ins and outs of networking on the Internet. Of course they use it every day, but because they work on the factory floor, terms like web services, HTTP and SOAP are sometimes foreign to them. I wanted to explain UA in a way so that these people could compare the technology to the kinds of technologies they already use and understand the applications and benefits of UA in industrial automation.

Lastly, I don't expect that you are going to do a cover-to-cover read of this book. It is designed to provide you with specific information on specific areas of OPC UA. For example, if there are terms you don't understand, you'll take a look at the dictonary sections. If you want to know how UA is being used, you'll take a look at the section on use cases.

Because I intend that you won't read it cover to cover, there is a lot of duplicate information in this book. I necessarily repeated myself so that you don't need to read the first section to get something out of the fifth section. Each section should stand alone, so if you're reading a section and thinking "I read this in the last section," this is the reason why.

The book was fun to write and hopefully valuable to you as a short, concise primer on UA. It will be up to you to determine how successful I have been.

John Rinaldi

8/12/2012

A LITTLE HISTORY OF OPC

Once upon a time, I gathered my engineering team in the RTA conference room. I set up my laptop and project. Put doughnuts on the table. I've found that doughnuts are a key ingredient to morning engineering meetings at our shop. And they had better be good doughnuts from a real bakery. Nothing else will do.

So I turned the projector on and OPC UA appears on the screen.

Now, some of them are conversant with OPC. They've heard the term. Heard me talk about it. Listened as customers described how they use OPC as a component of their factory automation architecture.

Others don't know it all. Don't know the history. Most are pretty young. None have real work experience on a machine startup. Six to eight weeks away from home. Living at the plant. Sweating out (sometimes in summer, really sweating) installation and troubleshooting of three or four thousand points, thousands of lines of ladder logic, plus configuration of all sorts of unique devices. And trying to make all of that electrical stuff work with the mechanical components of the machine.

So I started with the history of OPC. Back to the roots.

It was the 1990s. Very early in the development of the personal computer. Actually very early in the development of Windows. We're talking Windows 3.0, Windows 95, Windows for Work Groups.

This stuff was, to say the least, challenging to work with. Nothing integrated. Every application was a standalone application. Stuff crashed. Hard drives didn't last long. But that was okay because every nine months or so there was a new something to buy. New drives, new processor, new OS.

It was a time of unprecedented change. Impossible for anyone not working full time in the PC industry to keep track of.

In the midst of all that, people, usually systems integrators, were trying to incorporate these computers into factory floor applications. Looking back, it's kind of funny how they stood on their heads to integrate them. Special keyboard covers, cabinets, fans and filters. These were delicate machines, after all, with a lot of mechanical pieces.

And these brave souls even had to deal with labor issues. I recall that at some Procter & Gamble plants, unions demanded wage adders for anyone having to touch a computer keyboard. Increased salary came with a job where you worked with a computer.

They did all this to avoid what they considered overpriced programmable controllers. They figured they could write specialized applications to, if not replace those PLCs, bring a more sophisticated level of logic to these applications.

One of the things they incorporated was a lot of serial communications (long before Ethernet). PLCs were never very good with serial: not enough serial ports and lousy logic instructions to process the serial data. Any application that used a serial device was a candidate for a PC instead of a PLC.

But to implement these applications, the programmers had to write drivers for the serial devices. Need to get a flow rate out of flow meter? Or send the special control codes to the meter, receive and decode the response from the meter? Then you had to translate the binary data into a form that the application could use like real data or floating point.

Some of these drivers were pretty tricky. Years and years ago, I wrote serial drivers for lots of loop controllers. You had to send control codes to select a specific loop controller from the network (RS485 usually). Then you waited for a response. If a response didn't come, you had to take some action to let the application know or you processed the data.

Lots of times, you set up loops to run through a bunch of these. Sometimes, throughput or response time was crucial. Remember, we were doing this on old, slow PCs in the 1990s. A programmer could spend lots of time at this. Then there was troubleshooting in the field. Trying to figure out why only fourteen out of fifteen devices worked. Nightmare time.

And if you were a systems integrator, you were doing this over and over for every new project! Not a lot of those project managers slept well in those days.

But help was on the way. Around 1994-1995, a bunch of factory automation experts got together and decided there had to be a better way. Their work eventually led to the OPC 1.0 and the creation of the OPC Foundation.

Their mission was really pretty simple: create a way for applications to get at data inside an automation device without having to know anything about how that device works.

Pretty challenging mission. Nothing like this had ever been done before. How to make the thousands of automation devices accessible to any PC application.

Their solution centered on COM. COM was a fairly new a Microsoft technology at the time. It provided a way for Microsoft Windows applications to share data. One application could request data and another would supply it.

The OPC founding fathers took this technology mated it to an API that supported device protocols for automation devices, and OPC 1.0 was born. Figure 1 illustrates what this looked like when done.

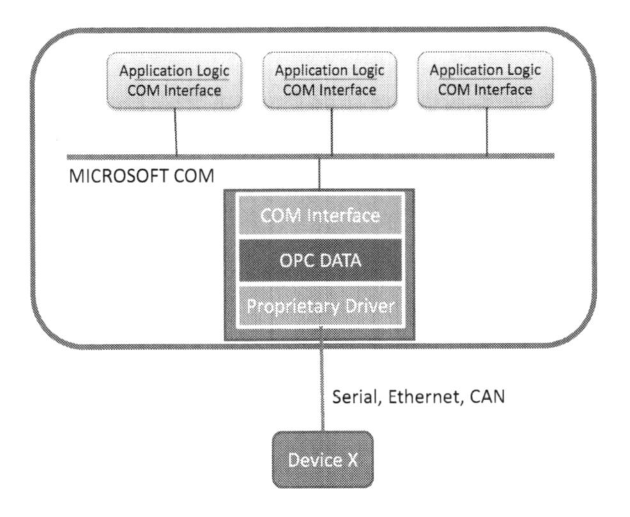

Figure 1 - COM Details

Data from the device, no matter what physical layer or protocol it supported, could be read or written by any Microsoft application that supported COM.

This was revolutionary and probably accelerated the adoption of PCs on the factory floor. For the first time ever, systems integrators and other factory floor application developers could implement applications without spending vast sums on driver development. Those project managers suddenly started sleeping a lot better.

And this technology made some people rich. Not Mark Zuckerberg rich, but "tollgate" position rich. Companies like Kepware, Matrikon and others owned the tollgate between your Microsoft application and all those automation devices. They began creating vast numbers of OPC servers that could be incorporated into all sorts of applications. The more OPC servers they created, the more applications people built.

But all wasn't well in OPC land. Soon, the applications demands and a changing automation landscape began to erode the acceptability of OPC servers in these applications. And that's what the next chapter in this saga is all about.

WHAT'S WRONG WITH MY OPC?

In the late 1990s and early 2000s, OPC spread like a weed. OPC servers were everywhere. Kepware and Matrikon and others deployed gazillions (well, thousands anyway) of OPC servers in every corner of the automation industry. Every type of industry. Every type of application.

But all wasn't that well with OPC or OPC Classic as I now call it [no relationship to the debacle known as Coke Classic]. Security issues, real and perceived, plagued OPC Classic.

Why?

People, mostly, and COM (DCOM is the distributed variant), the Microsoft technology underpinning OPC Classic, too. Let me explain.

COM is difficult to maintain and understand without significant training. And how you use it, configure it and set up authorizations varies slightly from one version of Microsoft Windows to the next. There are many ways to screw it up. And when you screw it up, your OPC Classic server stops transferring data and the people who want the data start screaming.

It's actually more insidious than that. If a well-meaning

but undertrained individual (not usually understanding that they are undertrained) goes in and fiddles with some COM or DCOM parameters, nothing happens. Literally, there is no affect. The OPC server keeps working.

For a while anyway.

Eventually, a week, a month or three months later, there's a reboot. Maintenance powered down the manufacturing cell, a year-end shutdown, whatever, but the machine is powered down. Now, when you bring the machine back up, all of sudden, for "no reason at all," DCOM stops working. The memory of that guy fiddling with the DCOM parameters is long forgotten. Instead, the OPC Classic server is blamed. It just stopped working.

It's a nightmare. The system needs to be fixed and another well-meaning, undertrained person starts fooling around. First thing they do: remove all the security. Generally, that makes it work. "Whew," they say. "Glad I could get that fixed."

Now you're set up for real trouble. There's a security hole in one of your servers that can lead a hacker down a path to who knows where.

And that's the kind of thing that makes the VP of IT lay in bed at night staring at the ceiling. He's the one whose butt is going to be called on the carpet if that hacker ever does attack.

And it does happen. Stuxnet is, of course, the classic case. Someone without malicious intent finds a "blank" USB stick and eventually plugs it into this now-unprotected server. Malware then starts looking for paths to specific automation equipment and all of a sudden you have a much bigger problem on your hands.

But in truth, the lack of DCOM knowledge and the seemingly inconsequential act of plugging in a data stick are really not OPC Classic problems. They're management issues. Management didn't dictate that a checklist be in place when an OPC Classic server stops communicating. Management didn't have a certification program in place ensure that the people maintaining OPC Classic servers were well trained in COM and DCOM and troubleshooting OPC Classic server problems.

It's just more convenient to blame the technology than ourselves and our management. So OPC Classic has taken some hits over the past few years in its public perception.

Probably not warranted, but what do they always say? "Perception is reality."

But there's another more pervasive problem with OPC Classic. One that can't be blamed on management. It's the deficiencies that come with dependency on Microsoft and Windows.

COM, the base technology for OPC Classic, is a Microsoft product. It runs only on Microsoft platforms. Not Linux, not VxWorks, not anything else. And that's a problem.

Microsoft has a well-deserved bad reputation, especially in industrial automation. In this industry, we generally build automation processes to last. There are a few products that are short-lived, but it's much more common to build production processes for diapers, soap, tea and hundreds of other products that we're going to run for the next five, ten or twenty years.

Microsoft products and PCs aren't suited for that kind of environment. Every time you buy a new laptop, you are hopelessly obsolete in, what, six months? How do you maintain an OPC Classic server on a Microsoft Windows platform for the next ten years?

So there's been a desire to run OPC Classic servers on other platforms. Platforms with longer lives and stable hardware that will last. Platforms that are smaller. Embedded platforms. The question being asked was "Why do I have to be tied to Microsoft? What about Linux? Why can't my flow meter and lots of other devices be an OPC server?

None of that is possible with Microsoft and DCOM.

There's also been dissatisfaction with OPC Classic in

the way that data gets to all those data-hungry servers in the upper echelons of the factory automation system and at the enterprise.

Most data is passed to those systems through a PLC. Possibly it starts in an RFID reader passing a pallet number, ID code and weight from the RFID reader to the PLC. From the PLC, it gets read by an OPC Classic server and passed to another application in the PC that transfers it to a logging database in the enterprise.

The real problem with OPC Classic is that this is an expensive and inefficient way to get data from a device (RFID reader) into that database. There's a PC involved – someone has to set it up, maintain it, validate that it is secure, etc. Initial hardware and labor and more ongoing labor.

But more than that, it's very inefficient and provides incomplete data. The data has to be carefully managed all the way from the RFID reader to the server to make sure that the different systems use the correct data types, that resolution is maintained, the endian order (which byte is first) is proper for that system. It's not easy.

And every time you decide you want a new piece of data, you have to touch multiple systems without breaking any of them. "Yuck" is the technical term for it.

Plus you don't get any of the meta-data. Meta-data being the associated data that provides the semantics for what you are transferring. Meta-data includes stuff like units, Scaling and all that other stuff that lets you work with the data without guessing as to what it is.

So, even though OPC Classic is wildly successful and works well when managed right, there was enough dissatisfaction with the Security issues, platform issues and data inconsistencies that a successor was planned for it.

And that's the subject of the next chapter.

WHAT IS OPC UA?

That is a very simple question. The answer when you are discussing a complex technology like OPC UA isn't as simple.

OPC UA, which I will refer to as UA throughout this article, is the next generation of OPC technology. UA is a more secure, open, reliable mechanism for transferring information between servers and clients. It provides more open transports, better security and a more complete information model than the original OPC, "OPC Classic." UA provides a very flexible and adaptable mechanism for moving data between enterprise-type systems and the kinds of controls, monitoring devices and sensors that interact with real world data.

Why a totally new communication architecture? OPC Classic is limited and not well suited for today's requirements to move data between enterprise/Internet systems and the systems that control real processes that generate and monitor live data. These limitations include:

- **Platform dependence on Microsoft** – OPC Classic is built around DCOM (Distribution

COM), an older communication technology that is being de-emphasized by Microsoft

- **Insufficient data models** – OPC Classic lacks the ability to adequately represent the kinds of data, information and relationships between data items and systems that are important in today's connected world
- **Inadequate security** – Microsoft and DCOM are perceived by many users to lack the kind of security needed in a connected world with sophisticated threats from viruses and malware.

UA is the first communication technology built specifically to live in that "no man's land" where data must traverse firewalls, specialized platforms and security barriers to arrive at a place where that data can be turned into information. UA is designed to connect databases, analytic tools, Enterprise Resource Planning (ERP) systems and other enterprise systems with real-world data from low-end controllers, sensors, actuators and monitoring devices that interact with real processes that control and generate real-world data.

UA uses scalable platforms, multiple security models, multiple transport layers and a sophisticated information model to allow the smallest dedicated controller to freely interact with complex, high-end server applications. UA can communicate anything from simple downtime status to massive amounts of highly complex plant-wide information.

UA is a sophisticated, scalable and flexible mechanism for establishing secure connections between clients and servers. Features of this unique technology include:

Scalability – UA is scalable and platform independent. It can be supported on high-end servers and on low-end sensors. UA uses discoverable profiles to include tiny embedded platforms as servers in a UA system.

A Flexible Address Space – The UA address space is organized around the concept of an object. Objects are entities that consist of variables and methods and provide a standard way for servers to transfer information to clients.

Common Transports and Encodings – UA uses standard transports and encodings to ensure that connectivity can be easily achieved in both the embedded and enterprise environments.

Security – UA implements a sophisticated security model that ensures the authentication of clients and servers, the authentication of users and the integrity of their communication.

Internet Capability – UA is fully capable of moving data over the Internet

A Robust Set of Services – UA provides a full suite of services for eventing, alarming, reading, writing, discovery and more.

Certified Interoperability – UA certifies profiles such that connectivity between a client and server using a defined profile can be guaranteed.

A Sophisticated Information Model – UA profiles more than just an object model. UA is designed to connect objects in such a way that true information can be shared between clients and servers.

Sophisticated Alarming and Event Management – UA provides a highly configurable mechanism for providing alarms and event notifications to interested clients. The alarming and event mechanisms go well beyond the standard change-in-value type alarming found in most protocols.

Integration with Standard Industry-Specific Data Models – The OPC Foundation is working with a number of industry trade groups to define specific information models for their industries and to support those information models within UA.

HOW OPC UA DIFFERS FROM PLANT FLOOR SYSTEMS

I've studied this technology for a long time now. And yet there is a question that I almost shrink from. In fact, I sometimes hate to answer it.

It's not because I don't understand what it is. It's not that I don't understand how it works. And it's not that I don't believe that it is a very valuable tool to almost every plant floor system.

It's just hard to put it into context when there isn't anything to compare it to. For example, when Profinet IO came out, I could tell people that it was the equivalent of EtherNet/IP for Siemens Controllers. Same kind of technology. Basically the same kind of functionality. Easy to explain.

But how do I explain UA when it doesn't have an equivalent? You could say that it's Web services for automation systems. Or that it's SOA for automation systems, an even more arcane term. SOA is "Service Oriented Architecture," basically the same thing as Web services. That's fine if you're an IT guy (or gal) and you understand those terms. You have some context.

But if you're a plant floor guy, it's likely that even though you use Web services (it the plumbing for the Internet) you don't know what that term means.

So the reason I get skittish about answering this question is that they always follow up with another question that makes me cringe: "Why do we need another protocol? Modbus TCP, EtherNet/IP and Profinet IO work just fine."

So I have to start with the fact that it's not like EtherNet/IP, Profinet IO or Modbus TCP. It's a completely new paradigm for plant floor communications. It's like trying to explain EtherNet/IP to a PLC programmer in 1982. With nothing to compare it to, it's impossible to understand.

That's where I am trying to explain OPC UA.

The people I'm trying to reach have lived with the PLC networking paradigm for so long that it's second nature. You have a PLC, it is a master kind of device and it moves data in and out of slave devices. It uses really simple, transaction-type messaging or some kind of connected messaging.

In either case, there is this buffer of output data in a thing called a programmable controller. There is a buffer of input data in a bunch of devices called servers, slaves or nodes. The buffer of input data moves to the programmable controller. The output data buffers move from the programmable controller to the devices. Repeat. Forever. Done.

That's really easy to wrap your mind around. Really easy to see how it fits into your manufacturing environment and really easy to architect.

OPC UA lives outside that paradigm. Well, really, that's not true. OPC UA lives in parallel with that paradigm. It doesn't replace it. It extends it. Adds on to it. Brings it new functionality and creates new use cases and drives new applications. In the end, it increases productivity, enhances quality and lowers costs by providing not only more data, but also information, and the right kind of information to the production, maintenance, and IT systems that need that information when they need it.

Pretty powerful, huh?

Our current mechanisms for moving plant floor data – few or no systems move information – is brittle. It takes massive amounts of human and computing resources to get anything done. And in the process we lose lots of important meta-data, we lose resolution and we create fragile systems that are nightmares to support.

And don't even ask about the security holes they create. Because when there are problems, and there always are, the first thing everyone does is to remove the security and reboot.

These systems are a fragile house of cards. They need to be knocked down.

And because of all this, opportunities to mine the factory floor for quality data, interrogate and build databases of maintenance data, feed dashboard-reporting systems, gather historical data and feed enterprise analytic systems are lost. Opportunities to improve maintenance procedures, reduce downtime, compare performance at various plants, lines and cells across the enterprise are all lost.

This is the gap that OPC UA fills. It's not something Profinet IO can do, even though the devoted acolytes would contest that statement. It's not something that EtherNet/IP can do. And it'd be a joke to talk about Modbus TCP in this context.

So I'm back to the original question: "What exactly is OPC UA"?

OPC UA is about reliably, securely and most of all, easily, modeling "objects" and making those objects available around the plant floor, to enterprise applications and throughout the corporation. The idea behind it is infinitely broader than anything most of us have ever thought about before.

And it all starts with an object. An object that could be as simple as a single piece of data or as sophisticated as a process, a system or an entire plant.

It might be a combination of data values, meta-data and relationships. Take a dual loop controller. The dual loop controller object would relate variables for the setpoints and actual values for each loop. Those variables would reference other variables that contain meta-data like the temperature units, high and low setpoints and text descriptions. The object might also make available subscriptions to get notifications on changes to the data values or the meta-data for that data value. A client accessing that one object can get as little data as it wants (single data value) or an extremely rich set of information that describes that controller and its operation in great detail.

OPC UA is, like its factory floor cousins, composed of a client and a server. The client device requests information. The server device provides it. But as we can see from the loop controller example, what the UA server does is much more sophisticated than what an EtherNet/IP, Modbus TCP or Profinet IO server does.

An OPC UA server models data, information, processes and systems as objects and presents those objects to clients in ways that are useful to vastly different types of client applications. And better yet, the UA server provides sophisticated services that the client can use, including:

Discovery Services – Services that clients can use to know what objects are available, how they are linked to other objects, what kind of data and what type is available, and what meta-data is available that can be used to organize, classify and describe those objects and values

Subscription Services – Services that the clients can use to identify what kind of data is available for notifications. Services that clients can use to decide how little, how much and when they wish to be notified about changes, not only to data values but to the meta-data and structure of objects

Query Services – Services that deliver bulk data to a client, like historical data for a data value

Node Services – Services that clients can use to create, delete and modify the structure of the data maintained by the server

Method Services – Services that the clients can use to make function calls associated with objects

Unlike the standard industrial protocols, an OPC UA server is a data engine that gathers information and presents it in ways that are useful to various types of OPC UA client devices, devices that could be located on the factory floor like an HMI, a proprietary control program like a recipe manager, or a database, dashboard or sophisticated analytics program that might be located on an enterprise server.

Even more interesting, this data is not necessarily limited to a single physical node. Objects can reference other objects, data variables, data types and more that exist in nodes off someplace else in the subnet or someplace else in the architecture or even someplace else on the Internet.

OPC UA organizes processes, systems, data and information in a way that is absolutely unique to the experience of the industrial automation industry. It is a unique tool that attacks a completely different problem than that solved by the EtherNet/IP, Modbus TCP and Profinet IO Ethernet protocols. UA is an information-modeling and delivery tool that provides access to that information to clients throughout the enterprise.

OPC UA DICTIONARY (PART 1)

One of the things that I've learned about OPC UA is that the terminology is a little different than what I'm used to seeing. With thirty years of industrial networking experience, I thought I had a lock on concepts like node and stack, but found myself unexpectedly confused when starting to study UA.

The terms used in lots of UA documents are similar to what I expected, but the designers twisted the meanings slightly. It's probably because UA is the first protocol that really crosses the line between the enterprise and the factory Floor. Because it has a foot in both worlds, the terms can be confusing to well-versed individuals in both the IT and factory floor worlds.

Another reason why the terms appear at first glance to be a little confusing is the scope of a UA discussion. In IA (industrial automation), we generally talk about interfaces between software components cohabiting in a processor. Or we talk about devices on the same subnet communicating over a very well-defined and very restrictive interfaces (EtherNet/IP, Modbus TCP or Profinet IO). In the Internet world, people talk about generic services with much more flexibility and capability than the interfaces between factory floor devices.

Here's part one of my dictionary. It covers general UA terms that I find interesting. (My apologies – It's not mean to be comprehensive.)

UA APPLICATION – In industrial networking, we generally draw a distinction between the end user application and the protocol stack. The end user application implements some set of defined functionality. The protocol stack moves well-defined data between the application and some external device using a very restrictive interface. Not quite the same in UA. In UA, I have found that the UA application references the end user application, the UA object model and the set of UA services implemented by the UA device. This is a much more encompassing use of the term.

UA CLIENT - A UA client endpoint is the side of a UA communication that initiates a communication session. Clients in UA are much more flexible than other network clients. UA Clients have the capability to search out and discover UA servers, discover how to communicate with the UA server, discover what capabilities the UA servers have and configure the UA server to deliver specific pieces of data when and how they want it. UA clients will generally support many different protocol mappings so that they can communicate with all different types of servers.

UA SERVER - A UA server endpoint is the side of a UA communication that provides data to a UA client. There is no standard UA server either in functionality, performance or device type. Devices from small sensors to massive chillers may be UA servers. Some servers may host just a couple of data points. Others might have thousands. Some UA servers may use mappings with high security and lower performance XML, while others may communicate without security using high performance UA Binary Encoding. Some servers may be completely configurable and offer the client the option to configure data model views, alarms and events. Others may be completely fixed.

BLOB (Binary Large Object Block) – "BLOBs" provide a way to transfer data that has no UA data definition. Normally, all UA data is referenced by some sort of data definition which explains the data format. BLOB data is used when the application wishes to transfer data that has no UA definition. BLOB data is user defined and could be anything: video, audio, data files or anything else.

PROTOCOL STACK OR STACK – A protocol stack like EtherNet/IP or Profinet IO in industrial networking generally implements the data model and the services of that protocol. An API connects that data model and service model to the data of the end user application. Though protocol stack vendors can implement this in many different ways, in general a UA protocol stack is comprised of three components: data encoding, security and network transport. Note that, unlike IA (industrial automation) protocol stacks, the data model and service model for the device are not necessarily included in the protocol stack.

ENCODINGS – A data encoding is a specific a way to convert an OPC request or response into a stream of bytes for transmission. Two encodings are currently supported in OPC UA: UA Binary and XML. UA Binary is a much more compact encoding with smaller messages, less buffer space and better performance. XML is a more generic encoding that is used in many enterprise systems.

XML is easier for enterprise servers to process, but requires more processing power, larger messages and more buffer space.

SECURITY PROTOCOL - A security protocol is the way to ensure the integrity and privacy of messages being transferred across a connection. UA uses the same type of security used on the Internet for privacy and security: certificates.

TRANSPORTS – A transport is the mechanism that moves a UA message between a client and server. This is another term that at first glance can be confusing. All UA messages are delivered over a TCP/IP connection. Within TCP, there is what I would call a session, though the word "session" is not specifically used anywhere that I've found in my study of the technology. There are two kinds of these sessions that message over TCP, and they are called transports when using UA. They are UA TCP and SOAP/HTTP. I discuss these in more detail in the next section of this dictionary.

MAPPINGS – This is an interesting term. The UA specifications are very abstract, unlike, say, a Modbus RTU specification. Modbus RTU runs over multi-drop RS485 and that fact is inherent in the specifications. It's not that way with UA. The specifications for UA operation are very abstract and done that way to maintain the ability to take advantage of future technologies. A mapping refers to how those abstract specifications are mapped onto a specific technology. For example, a security mapping describes how the UA Secure Channel Layer is implemented using WS Secure Conversation. A UA Binary Encoding mapping describes one way that UA data structures are mapped into a stream of bytes.

API (Application Program Interface) – An API is the

set of the software interfaces that allow one software application to use the services of another software application. In the industrial world, this normally refers to the interface between two pieces of software cohabitating in the same processor. In the Ethernet world, the API can refer to the interfaces needed by some client device to access the services available of some remote web service. In UA the API generally refers to the set of interfaces that a UA toolkit vendor provides to a device developer. Because different toolkits are designed differently, the APIs work differently. The API may include interfaces to the data model. In other cases, the API may only interface the three main components of UA: the encoding layer, the security layer and the transport layer. The UA data and service model may be part of the user application.

OPC UA DICTIONARY (PART 2)

This is the second part of the dictionary I've been building as I study OPC UA technology. The terms in this section refer to items that are more pertinent to developers of OPC UA stacks though could be interesting to those who desire a deep understanding of OPC UA technology.

The list is not inclusive of every term that applies to OPC UA stack technology.

WEB SERVICES – Web services is a generic term for loosely coupling Internet services (applications) in a structured way. The majority of Internet applications are today built using Web services. With Web services, you can easily find services, obtain the interfaces and characteristics of the interfaces and then bind to them. HTTP, SOAP, XML are the basic technologies of Web services applications and are some of the technologies that can be used by OPC UA clients and servers.

SERIALIZATION – This is an easy term to

comprehend. This is the process of taking a service like the read attribute service and creating the series of bytes that a UA server can process and return the value of an attribute. Serialization dictates how data elements like a floating point value are transformed into a series of bytes that can be sent serially over a wire. Two types of serial encoding are currently supported by UA: UA Binary and UA XML.

UA XML ENCODING – XML encoding is a way to serialize data using Extensible Markup Language (XML). An encoding is a specific way of mapping a data type to the actual data that appears on the wire. In XML encoding, data is mapped to the highly-structured, ASCII character representation used by XML. XML can be cumbersome, large and inhibit performance, but the encoding is used because a large number of enterprise application programs support XML by default.

UA BINARY ENCODING – UA Binary encoding is a way to serialize data using an IEEE binary encoding standard. An encoding is a specific way of mapping a data type to the actual data that appears on the wire. In Binary encoding, data is mapped to a very compact binary data representations that use fewer bytes and are more efficient to transfer and process by embedded systems. Binary encoding is widely used by industrial automation systems, but less common among enterprise applications.

SECURITY PROTOCOL – A security protocol protects the privacy and integrity of messages. OPC UA takes advantage of several standard, well-known security protocols. The selected security protocol for a specific application is a combination of the security requirements for the installation and the encoding and transports selected for OPC UA implementation.

TRANSPORT PROTOCOL – A transport protocol (also referred to as a "transport") provides the end-to-end transfer of UA messages between UA clients and servers. Once a UA service message is encoded and passes through securitization, it is ready for transport. Two transports are currently defined for OPC UA: UA TCP and SOAP/HTTP. The underlying technology for both these transports is standard TCP. TCP provides the socket-level communication between clients and servers.

UA TCP TRANSPORT – UA TCP transport is essentially a small protocol that establishes a low-level communication channel between a client and a server. Most of what the UA TCP transport does is to negotiate maximum buffer sizes so both sides understand the limits of the other. The advantage of UA TCP is its size and negligible impact on throughput.

HTTP (Hypertext Transfer Protocol) – is part of the basic plumbing of the Internet. It is the low-level protocol that allows a client application like your browser to request a web page from a web server. HTTP messages request data or send data in a very standard format supported by every Internet-aware application.

XML (Extensible Markup Language) – XML is a highly structured way of specifying data such that applications can easily communicate. XML transfers all data as ASCII – the one commonly understood data format for all computer systems. XML uses a grammar to define the specific data tags that are used by an application to pass data.

SOAP (Simple Object Access Protocol) – SOAP extends XML and provides a higher level of functionality. Among other things, SOAP adds the ability to make remote procedure calls within an XML structure.

HTTP/SOAP UA TRANSPORT – The HTTP/SOAP transport is the second transport currently supported in OPC UA. This transport requires larger messages, bigger buffers and more processing, but is used because HTTP and SOAP are supported by almost all (if not all) enterprise applications. It is a standard way of moving serialized OPC UA messages between a client and a server.

OPC UA SYSTEM CONCEPTS

This is a section that a lot of readers will find pretty straightforward and comfortable. I'm going to discuss UA from a system point of view. We'll take a look at the architecture as a whole and then dive into the individual components.

UA uses a Web services kind of architecture. There is another section in this document that discusses Web services in detail. For this section, we should know that Web services is an architecture where systems are de-coupled to provide faster, more flexible interaction between disparate systems. And that's exactly the goal of UA: to be flexible enough to link major business systems like SAP with small embedded devices such as a motor controller. A key component of the Web services architecture is that you can easily find the available services, obtain the interfaces and characteristics of the interfaces and then bind to them. This is the architecture that has made the Internet so successful.

Web Services and UA both use a client – server type architecture. Client devices seek out, discover, interrogate, connect to and configure one or more server devices.

Server devices provide endpoints for specific interactions and make those endpoints available to client devices. A device called a discovery server assists clients in finding servers. Let's look at each of these components of the UA architecture in more detail, beginning with server devices.

Server devices encapsulate data and information using the UA address space. There is another section in this document that describes the UA address space and its object-oriented structure, but let's start with the structure shown in figure 1.

OPC UA SERVERS

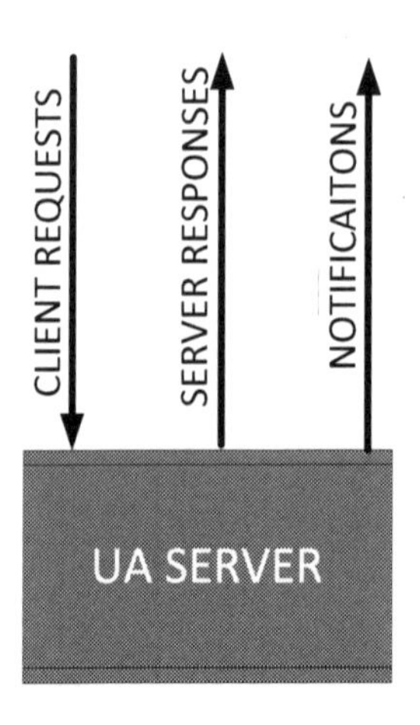

Figure 1 - External View of the Functionality of a UA Server

From an external perspective, a UA server simply receives client requests, provides responses to the UA client and publishes notifications. Notifications inform clients about events to which they have previously subscribed.

What kinds of client requests does the server handle? OPC UA servers provide a long and varied list of services to OPC UA client devices. Some of the more important ones include:

Discovery – OPC UA servers have to make themselves available to be discovered by UA clients. They offer a discovery endpoint that can be found directly or through a discovery server.

Profile Support – UA servers provide their profile information to clients so that the client devices can determine if the server has the capability to meet the requirement of the client application. For example, a client may require a server to support a security level that the server is unable to meet. In that case, the client would not make the connection.

Address Space – The most basic service that a UA server provides is to allow the client access to the data encapsulated in the address space of the server. The client must be able to view the address space, read properties of the nodes it finds in the address space and read and write attributes of the variable type nodes in the address space

Notification/Subscription Support – Servers that support this functionality allow a client to define a set of nodes that the server will monitor for some condition and subscribe to notifications when that condition is detected.

Views – Servers provide the capabilities for the clients to group specific nodes in the address space of the server as a view.

The architecture of a server is illustrated in Figure 2.

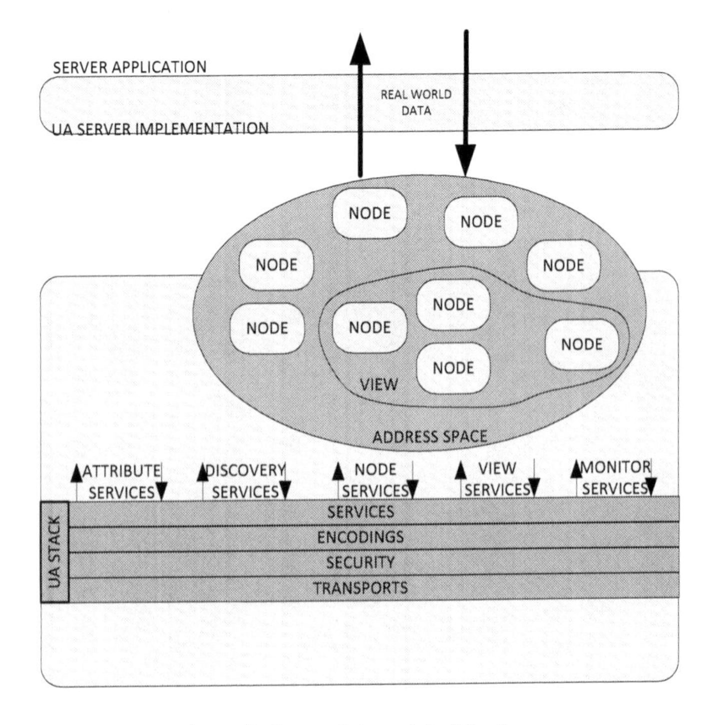

Figure 2 - Server Internal Architecture

Let's take a look at each of the parts of a UA server.

The most basic part is the UA stack itself. This is where the heavy lifting of moving messages happens. For outgoing messages, this is where the outgoing message is converted into a stream of bytes (encoding), the message and connection is authorized and the low-level transport of the message from point A to point B happens. For messages coming from a client, that all happens in reverse.

This part of the server implementation can vary with the specifics of the implementation. For example, a device can support both XML SOAP-HTTP encoding/transport and UA Binary-UA TCP encoding/transport. The endpoint that is used by a client depends on the environment in which the UA server is located. On the manufacturing floor, the device may use UA Binary-UA TCP, but on the business systems network it may use the XML SOAP-HTTP.

But that's just moving messages. There's a lot more important stuff that goes on. For example, a message from the client must be decoded and acted upon. That is in the services area of the server. Services are shown in the figure 2 diagram as part of the stack, but also as part of the UA implementation, as one might claim it to belong in either area.

A service request like "read attribute" is processed in this section of the server. Processing of the request generates a response message, which is passed through the communications part of the stack as described above. That happens for all service requests. The client request is processed, a response created and then it is encoded, securitized and transported to the client.

The UA implementation is the part of the server that implements the UA address space. This is where the application synchronizes the real world data to the UA address space. Sensor or other data collected from the real world is stored as objects in the UA address space. Output data in the address space is written to the real world outputs. And in more sophisticated applications, this data is organized to conform to a specific information model.

The entire UA implementation, services and everything else can be thought of as part of the UA stack, but I like to separate the UA implementation from the UA communications. They really are two separate items, though if you buy a UA stack you will find that there is no clear delineation between the two.

Lastly, the final part of a UA server is the server application. If the device is a motor controller or some other sort of device, this is where the real work of the application happens.

OPC UA CLIENTS

At this point you may be asking, "What about the Client?" Well, I haven't forgotten about that. It's just that clients are all over the place. Clients can be part of a database application, a small HMI or anything in between. They can be on powerful servers and have full functionality, or be a simple client that just reads and writes a couple of attributes. They are much harder to characterize than UA servers, so there isn't a whole lot to discuss.

Client functionality is really application-dependent, though there are a few components that all clients have in common. One, clients are able to find servers using the discovery service. This is the most basic service, as nothing else can happen unless they can find a server. Second, clients are able to interrogate the server to identify the profile being supported and determine if the server can meet the needs of its application. The client must determine if the server supports the services, security level and functionality that is required. Thirdly, the client must be able to use the services supported by the server to read/write attributes, set notifications and do all the things that are required to accomplish its application task.

OPC DISCOVERY SERVER

The last part of the architecture of UA is the discovery server. This is a special type of server that only exists to sort of catalog the servers that are on the network. It is a place where a client can go to find the available servers without doing a lot of searching. The usefulness and viability of this kind of server for the majority of applications is still in some doubt.

THE OPC UA ADDRESS SPACE MODEL

I know a number of people that talk in riddles. As much as I try, I can't seem to fathom what they are trying to communicate. No context. No patterns. No structure. Just data dumps right from their head.

Data, in and of itself, is often like that. Not much use to anyone. Placing structure, patterns and context around raw data is the definition of an address space model. In other networks, it's called the data model, the data representation, the object model – there's a whole host of names. No matter what the name, an address space model specifies a mechanism for organizing and addressing data.

OPC UA is built around a very highly detailed, superbly organized address space model. The UA model is more sophisticated than object models of the industrial protocols (EtherNet/IP, Profinet IO and Modbus) and the building automation protocols (BACnet and LonWorks).

For the most part, the industrial protocols and the building automation protocols all use a straightforward and well-defined structure that encapsulates the data as objects. Some have other types of organizations: Lon uses network variables, and Modbus uses registers and coils.

These models are adequate for the space they occupy in the automation hierarchy (I/O), but they are pretty inflexible and inadequate for general-purpose data transfer. The structures are usually predefined and fixed. They transfer data but provide little to no access to meta-data (scaling, engineering units, etc.). There is usually no ability to dynamically alter the available data sets, make new relationships between objects, or dynamically create event notifications or new objects.

OPC UA provides all this and more. We'll start by taking a look at the central concepts for UA, the object and the node.

An OPC UA server exists to model data, information, processes and systems and present those items as objects to clients in ways that are useful to vastly different types of client applications. Objects can be anything from a tiny sensor to an entire production facility and everything in between. Because UA is not designed to operate in one specific domain, UA servers must be have the flexibility to present that data and information to clients in various ways that are useful to vastly different kinds of client applications.

Every UA object is a node. A node is the central structural element of the UA address space. Objects and everything else in UA are built on the node concept.

All nodes use an identical structure. Nodes are described by their attributes and are interconnected by references to other nodes.

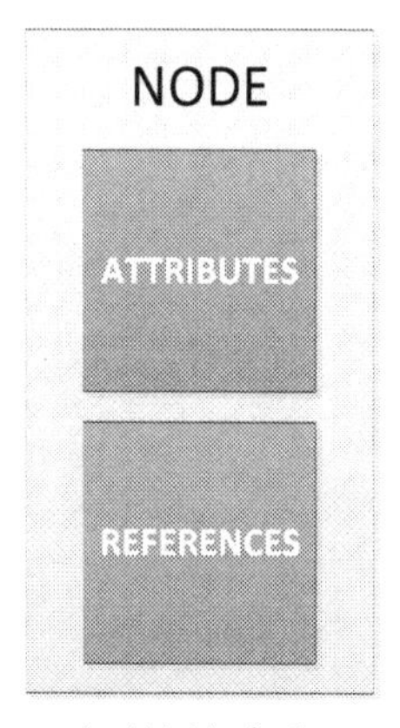

Figure 1 - UA Node Structure

Nodes are all derived from the base node class plus one of the eight node classes defined in OPC UA. The eight node classes of OPC UA include:

- Objects – Representation for real world data, processes, software objects and other items
- Variables – Nodes that contain values
- Methods – Nodes that encapsulate some programmable logic
- Object Types – Type definitions for the object nodes
- Variable Types – Type definitions for the variable nodes
- Reference Types – Type definitions for the references that connect one node to another
- Data Types – Actual supported data types
- View – A limited area of interest within a larger address space

Objects, for example, are derived from the base node class plus the object node class. The base node class defines the structure common to all nodes, while the object node class defines the structure common to objects in an OPC UA system.

Nodes with the object node class are the organizing mechanism for the OPC UA address space. Objects through their references to a group of nodes define a sensor, process, production line or plant. Objects, like all other nodes, consist of attributes and references.

Attributes characterize the node. Attributes describe the names used, the node information, the data type and the value present in an object. Attributes are part of a node and are not directly visible (not browseable by a client) in the address space. There are several types of attributes:

- Base node attributes are inherited from the base node type definition and provide the identical characterization as all other nodes.
- Node class attributes are inherited from the type class for the node type. Attributes of an object node inherit the attributes of the object node class type. Attributes of variable nodes inherit the attributes of the variable node class type.

References are links to nodes associated with a node. Objects of the object node class have variables (variable node class) and methods (method node class) associated with them. References link a source node (an object in this example) to a target (one or more variable and/or method nodes). Methods are stateless functions that can be executed by the client process.

The address space of a UA server is composed of a series of objects that it makes available to UA clients. Each object references a group of nodes that characterize the information, process and system represented by the address space.

Variables are nodes derived from the variable type class and deserve special mention here as they are the only nodes that can contain values. Variables have the identical structure of all other UA nodes. There are two kinds of variables: data variables and properties.

- Data variables are nodes of type variable class containing values. These values pertain to the specific functionality of the object. An oven temperature, speed, unit count, purchase order number and engineering work order number are all values that would be found in data variables.
- Properties are nodes of type variable class that provide semantic information for an object. A property could be the engineering units, the previous order date for a purchase order, the operator name or other non-process related data

THE OPC UA INFORMATION MODEL

Many automation engineers and even a lot of the people on my staff are confused by the term "Information model." It's not a term that they have encountered before because it doesn't really exist in the networking technologies used in their factories.

An information model is a structure imposed on top of the address space that relates pieces of the address space together in such a way that these elements represent standard systems, processes, devices, manufacturing cells or entire plants.

There are two kinds of information models. There are non-standard models created by integrators, vendors or other professionals to standardize the interfaces of a proprietary or local installation. And there are standardized models that provide common interfaces to entire industries. For example, there are information models that represent the IEC 61131-3 architecture and work underway to provide standard representation for the oil and gas industry.

Let's look at a very simple example of the first type. In

this example, an installation has a number of tanks with simple pumps and level sensors. It's a very simple system. When the level falls below a set point, the pump runs until the level is restored.

For our purposes, we will assume just two variables. There is an analog level indication and a pump with a discrete output that indicates it's running or not. Any standard protocol can encapsulate the data of this system: Modbus, EtherNet/IP, Profinet IO, BACnet and others. In Modbus, there is simply an input coil and an input register. It's equally simple in any other protocol, but it's different in UA.

In UA, the designer of the information model defines a pump as one object and a level sensor as another object. Each of these objects contains not only the data values, but also meta-data that describes the data type for each of the variables of these objects. And there is a reference from the level sensor object to the pump object as to how they are related.

And to encapsulate the entire system, there is a higher-level object that references both of them. We'll call that the tank object. The tank object is the system object that links all the elements of the system together. Meta-data for the tank object can describe where the tank is located, the size of the tank and more.

The beauty of this Information model is that a client device can first interrogate the objects of the server device, find the tank object and then from the tank object follow references to all the components of the object. The client can discover the meta-data for the system (the tank) or any of the component objects that have data.

And it gets even better. If there are multiple tank objects, the entire structure can be duplicated and a higher-level object (in this case, a tank farm object) can be created to support a higher-level system. This is truly information now, not just a coil and a register like you would have in Modbus or some other simple protocol.

This kind of hierarchical representation of information is a true Information model built upon the standard address space model of UA.

THE OPC UA SECURITY MODEL

Note: OPC UA Security is only one part of an overall security plan for an installation. As OPC UA is a communications protocol, the security focus is on communications security: the integrity of messages between clients and servers. Developers must use the UA security model as one component of the overall security plan.

Security was easy in the old days. There just was no connectivity between anything on the factory floor and the Internet. That's the famous "air gap." A hacker would have to physically penetrate the walls, guards and access systems to create any mischief on the factory floor.

That was fine for those days, but the luxury of maintaining that air gap is gone. Now MIS systems, MES, ERP and other systems must have access to factory floor controllers and sensors to achieve the productivity, quality and interconnectedness required of 21st century automation environments. Today's manufacturing plants are not only physically larger, but also more geographically diverse and require a level of coordination and integration unfathomable just a few years ago. Integrated systems spanning countries, suppliers and technologies are common. The Internet has become the vital infrastructure for this strategy, and with it, a huge risk to the enterprise.

There are lots of groups and individuals ready, willing and able to launch attacks against automation systems. Attackers to be concerned about include hobbyists, professional hackers, malware, nation states and disgruntled employees. Of these attackers, malware and employees are the easiest to defend against. Hackers who can quickly modify their strategy after sophisticated probing of defenses are of the most concern.

Attacks can come in many forms, from phishing, to denial of service, to man-in-the-middle, to memory injection and program modification. These kinds of attacks can be leveled at any type of manufacturing floor device, including controllers, actuators, HMIs and more. Since most control engineers still optimize systems for availability, preventing these attacks has not been a priority, leaving many of systems at risk.

Knowing that OPC UA would often be used to cross the gap between the factory floor and the enterprise and that UA would be used in a diverse range of applications, security was an important component in its design. There are four goals to OPC UA security:

1. Providing a mechanism for authentication of clients and servers – verifying that each side knows it is talking to an authorized device
2. Authenticating users – verifying that users configuring clients and servers are authorized to make revisions to the operation of the clients and servers
3. Integrity and authenticity of communications – ensuring that messages cannot be intercepted and decoded

4. Validating claims of authenticity

The OPC UA security model is flexible model in which security measures can be selected and configured to meet the security needs of a given installation and application. Obviously, the security requirements for a water purification plant might be different than a small, highly self-sufficient crayon manufacturer. OPC UA is flexible in order to meet the needs of applications with varying security needs.

Security – the threats UA is designed to mitigate and how UA mitigates those threats – is a huge topic, much too large for discussion in this short book. Here is a summary of the relevant components of OPC UA security:

Certificate-based authentication - A way of using encryption to positively identify what computer and/or user is making a request. By knowing the requester, "man in the middle" attacks can be prevented and all requests from non-authorized sources can be blocked. Certificate-based authentication is much more secure than common practice, which is to accept any computer presenting the correct address and computer name as that computer.

Authorization – UA is designed to work within the overall security authorization infrastructure employed by the facility. There is no requirement about how those authorizations are managed.

Security Audit Trails – Audit trails provide traceability between client and server activities. If a security-related incident is discovered at the server, the client audit log for that event can be located and examined. Servers can implement event notifications to report auditable events to clients capable of processing and logging them.

Message Confidentiality - UA uses encryption to protect the confidentiality of messages between a client and a server

Integrity – UA uses symmetric and asymmetric signatures to address integrity as a security objective.

Transport Security - Three security facets are provided, with three levels of transport security. Transport security and secure channels can be implemented for either of the available UA transports: HTTP-SOAP or UA TCP transports.

Discoverable Security Configuration – The server discovery mechanism provides a way for clients to detect the security configuration of a server and provide the certificates and/or transport security required for that server.

OPC UA PROFILES AND FACETS

OPC UA is without a doubt the most scalable networking technology that I have encountered in my twenty-five years in automation. But, of course, the designers of the technology had no choice. How else do you build something that functions on the most advanced, high-end server environments using state of the art tools and in a simple sensor that simply needs to send a humidity reading every hour?

Profiles, facets and conformance units are all concepts that support the scalability inherent in UA. Let's start by defining each of these terms:

Conformance Units – Conformance units are simply testable units of functionality. The call service is an easy to understand example of a conformance unit. One or more conformance units define a profile. Conformance units are not specific to a particular profile. You will find the same conformance unit in multiple profiles.

Profile – a set of functionality that defines the capabilities of a UA device. A profile indicates to other devices (electronically) and to people (human readable form) what specific features are supported. Engineers can determine from the profile if this device can be used in an application. A client device can determine if a device has the functionality required for the application and if it should initiate the connection process with the device. There are two types of profiles: full-featured and facets. Full-featured profiles define a set of conformance units that a group of applications needs to support. The nano-embedded server profile contains the minimum set of conformance units needed by small embedded devices. All devices must be based on a full-featured profile.

Facets – Facets are profiles that contain specific functionality (conformance units). Facets are added to a full-featured profile to extend the functionality of the profile with that additional functionality.

Profiles are further grouped into categories related to specific user requirements. There are server profiles, client profiles, transport profiles and security profiles. The profiles within each of these groups provide different kinds of functionality within each area.

The HTTP SOAP transport profile, for example, defines the functionality needed to transport messages between high-end Internet-type devices. Devices that support that transport profile may not need to support the UA TCP protocol. Selection of a particular profile doesn't exclude inclusion of an alternative set of functionality in the device.

Security profiles are worth a special note. Not many embedded devices have the hardware and software platforms that can manage the encryption of messages in a timely manner. But some do. So the security profiles are facets that can be added to the standard security profiles.

Clients by nature are very diverse devices. There are no general categories for client devices and therefore no specific set of client profiles. Some clients will use XML encoding while others will use both XML and UA Binary.

There are no specific set of features that can be easily grouped.

Profiles are the basis for the OPC Foundation Certification Process. Certified devices are tested based on the Profiles supported and the Conformance Test Units embedded in the device.

If pass conformance you are able to use the official OPC Foundation Gold Test logo. Conformance testing is free to members of the OPC Foundation:

OPC UA CASE STUDY

In this section I'd like to illustrate the power of UA with a simple example: automatic door systems. These are familiar systems found all over the world. You find them at shopping centers, airports, drug stores and many, many other places.

These systems can be rather complex, but for our purposes we are going to limit them to a motor, two sensors and an Ethernet-enabled controller of some sort. An actual system might include a people counter, lighting controls, temperature sensors and more, but we'll keep it simple in this example.

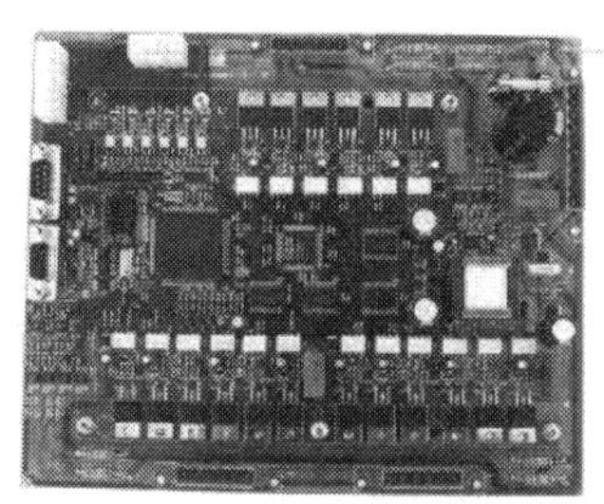

Figure 1 - Automatic Door Components

In our example, the automated door system has only two electro-mechanical requirements:

1. Enable and disable operation of the door based on the facility schedule
2. When in the enabled mode, open the door when indicated by the sensor input

In practice, if the manufacturer chose to connect the system to the back office systems, they could add any number of new features and benefits:

1. Track the number of activations and automatically schedule preventive maintenance
2. Record the number and time for all after hours override door activations

3. Detect motor, sensor and control board failures remotely
4. Detect operational failures
5. Easily update the operating schedule remotely
6. Detect and report unauthorized intrusion
7. Monitor customer entrance and exit activity

And on and on. Once this simple system is connected to the back office, there are many, many possibilities for collecting all sorts of useful data that can be turned into information.

So, if the vendor of this system chose to connect this controller to the back office, what are some of the options that might be considered?

OPC UA offers is a great choice for this application because of the inherent advantages in the technology.

1. Secure access – UA is the only network that can provide access to these automation door systems across the country to collect operating data in a secure way. There is no other network that is both capable of crossing a firewall and traveling over the Internet with this level secure way.
2. Alarming and event notifications – UA clients can ask each server to notify it when certain attributes exceed a limit. This means that notification will happen when maintenance is due.
3. Customized object model – The automated door vendor can layer an information model on top of the UA address space and make access to every door attribute available from one object reference.

4. Customized views – The UA client can access any segment of data within the information model as needed.
5. OPC UA extensions for databases and other tools – These interfaces allow databases like Oracle and Sequel to easily collect data from the automatic door systems with little or no integration.
6. Automatic discovery – The automatic door vendor can employ a discovery tool on the Intranet or the Internet to provide a "catalog" of all the automatic door vendors.
7. Audit trails – Any unauthorized attempt at access to a server results in an audit log entry and notification of the UA client that an intrusion has been detected.
8. Corporate access – A facility with lots of automatic doors in diverse geographic locations can easily change door schedules from a central location.
9. Standard Interface – HMIs, PC Tools and other software can easily interface the open UA communications and access status data and diagnostics on doors.
10. Customers buying the automatic door system can use the Open UA Interface to develop their own management tools for their systems.

There are several ways to implement UA on this automatic door controller. That is the subject of a later chapter of this book.

LOOSE VS TIGHTLY COUPLED CONTROL SYSTEMS

There's a lot of talk today about the integration of the enterprise and the factory floor. I've enjoyed a lot of this discussion. Rockwell has a great image of something they call the "Convergence Man." It's a guy split in two with a hardhat and factory smock on his right side and the more formal shirt and pants of an IT guy on his left side.

There's also a bunch of new terms being thrown around like "digital factory" and "integrated intelligence," a sign of the increasing talk (and action) toward linking the factory floor with systems not directly involved in factory floor control. Enterprise systems can be big and sophisticated like ERP and MES systems, or they can be as simple as a recipe manager on your server that downloads twenty tags once a day.

No matter what you're doing, there's a key distinction between the systems on the factory floor and in the enterprise that not many people understand. This difference involves what I call "loosely-coupled" systems and "tightly-coupled" systems. I don't think these are new concepts, but I don't think they've been examined in the light of the current trend toward the integration of factory floor and enterprise systems.

I would argue that factory floor systems should be labeled "tightly-coupled." Systems that use Profibus, Profinet IO, DeviceNet, EtherNet/IP or any Modbus version have a very strict architecture. These are really I/O systems, despite that the folks at the ODVA (the Open Device Vendor Association) and PI (Profinet International) would have you believe otherwise.

Let's look at the main characteristics of these tightly-coupled systems:

A Strictly Defined Communication Model - The communications between these systems is inflexible, tightly regulated and as deterministic as the communication platforms allow.

A Strictly Defined Data Model – The data (really I/O for most of these systems) model is predefined, limited and inflexible.

Strictly Defined Data Types – The data types transported by these systems are limited, predefined and supported by both sides. There is no ability to send data in an open and universal format.

We could look at any of the factory floor protocols, but let's take EtherNet/IP as an example. EtherNet/IP has a very strictly defined communication model. A scanner uses a very precise communications model in communicating with its adapters. The adapters are preconfigured, all data exchanged is predefined and nothing changes without human intervention. The data exchanged is part of the adapters predefined object model and the data is formatted in a way supported by both the scanner and the adapter.

Tightly-coupled systems provide much needed, well-defined functionality in a highly specific domain. Expanding operation to other domains or trying to provide more general operation is difficult. Making more generic data and functionality available requires significant programming resources that results in a very inflexible interface.

And that's why tightly-coupled systems are wrong for enterprise communications. That is why I continue to be amused by the proponents of EtherNet/IP and Profinet IO who promote them as ways to exchange data with enterprise systems. Can they be made to work for a specific application? Yes. But to get there requires a whole lot of effort and results in a difficult-to-maintain, inflexible system that is extremely fragile.

Loosely-coupled systems, on the other hand, provide exactly the right kind of interface for enterprise communications. Loosely-coupled systems decouple the platform from the data, the data from the data model and provide a much more dynamic mechanism for moving data.

Loosely-coupled systems have these kinds of characteristics:

A Widely Used, Standards Based Transport Layer - Messages are transported in loosely coupled systems with open, widely implemented, highly flexible transports layers: TCP and HTTP.

An Open, Platform-Independent Data Encoding – Data is encoded using an open standard data encoding like Extensible Markup Language (XML) that can be processed by any computer platform.

A Highly Extensible Operating Interface – The interface between loosely-coupled systems is flexible and extensible. SOAP (Simple Object Access Protocol) is the main interface and it provides a highly flexible mechanism for messaging between loosely-coupled systems.

Essentially what I've described here is Web services. Web services is the backbone of everything we do on the Internet. It is extensible, flexible and platform-independent – all required for maximum functionality on the ever-expanding Internet.

The challenge is to how to best connect the tightly-coupled factory floor architectures with the loosely-coupled Web services architecture of the Internet. Rockwell has its Factory Talk product line. The ODVA promotes EtherNet/IP. PI promotes Profinet IO.

Any of these can be made to work as I have described earlier, but the EtherNet/IP and Profinet IO type protocols result in that dreaded brittleness that costs too much time and money over the long term. These approaches take massive amounts of human and computing resources to get anything done. And in the process, we lose lots of important meta-data, we lose resolution and we create fragile systems that are nightmares to support.

And don't even ask about the security holes they create. These systems were not designed to be highly secure.

These systems are a fragile house of cards. They need to be knocked down.

And because of the discontinuity between the factory floor and the enterprise system, opportunities to mine the factory floor for quality data, interrogate and build databases of maintenance data, feed dashboard reporting systems, gather historical data and feed enterprise analytic systems are lost. Opportunities to improve maintenance procedures, reduce downtime and compare performance at various plants, lines and cells across the enterprise are all lost.

The solution? I've thought a lot about it and I think it's OPC UA because UA can live in both the world of the factory floor and the enterprise.

OPC UA is about reliably, securely and – most of all – easily modeling "objects" and making those objects available around the plant floor, to enterprise applications and throughout the corporation. The idea behind it is infinitely broader than anything most of us have ever thought about before.

And it all starts with an object. An object that could be as simple as a single piece of data or as sophisticated as a process, a system or an entire plant.

It might be a combination of data values, meta-data and relationships. Take a dual loop controller: the dual loop controller object would relate variables for the setpoints and actual values for each loop. Those variables would reference other variables that contain meta-data like the temperature units, high and low setpoints and text descriptions. The object might also make available subscriptions to get notifications on changes to the data values or the meta-data for that data value. A client accessing that one object can get as little data as it wants (single data value) or an extremely rich set of information that describes that controller and its operation in great detail.

OPC UA is, like its factory floor cousins, composed of a client and a server. The client device requests information. The server device provides it. But as we see from the loop controller example, what the UA server does is much more sophisticated than what an EtherNet/IP, Modbus TCP or Profinet IO server does.

An OPC UA server models data, information, processes and systems as objects and presents those objects to clients in ways that are useful to vastly different types of client applications. And better yet, the UA server provides sophisticated services that the client can use, like the discovery service.

I think that UA is the future and the perfect technology to bridge the chasm between loosely- and tightly-coupled systems.

OPC UA AND WEB SERVICES

Why Are We Talking About Web Services?

I've included this chapter in this little book because a lot of industrial automation people aren't really all that literate with the plumbing of the Internet. And in the 21st century, the Internet and all that technology is steadily working its way into the manufacturing/automation environment. The more we know about it, the better off we will be to meet the challenges in the future.

So you've heard the term Web services. But what does it really mean?

Let's start by thinking about when the Internet was new. Moving data between two systems was difficult. Look at the challenges just to move something simple like a bank balance. How do you represent $1,000? Is that two bytes of information? Or is it four bytes of information? Because if you save your money, that $1,000 may exceed $65,535 someday and you can't represent that in two bytes. If it's four bytes, is the first byte the most significant, or is it the last? What about the decimal places when I deposit ten more cents? Lots of platforms and processors have radically different mechanisms for storing that simple piece of data. What a mess!

But they got past that pretty darn quick. Somebody decided that all data sent between two platforms should be sent as ASCII. Every system does ASCII in the same way. $1,000.10 is sent as nine bytes of data from the "$" to the final "0." And to make sure that everybody understood what data what being sent, they created a structure called Extensible Markup Language (XML) to pass information about the data. Now the nine characters of data became "<BANKBALANCE>$1,000.10<\BANKBALANCE>." The initial "<BANK BALANCE>" is the opening tag for the amount, while the "<\BANK BALANCE>" is the closing of the tag.

Great, but what if I want to know how many Euros I could buy right at this moment with that $1,000? Now I need to pass that dollar amount to another system and ask it to execute some logic on that number. That's a much bigger problem.

Initially, this kind of computer-to-computer integration was all hard coded. An army of programmers would attack it, define some kind of interface between the two systems and program both sides so that the request could be sent, received, processed and responded to. And then when we wanted to do a Yen to Euro conversion, the army of programmers would jump into action again. All this programming was expensive. Lots of Labor. Lots of time. Lots of money. Yuck!

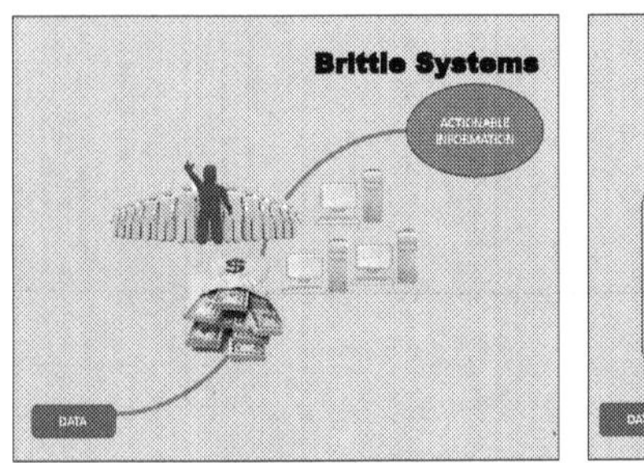

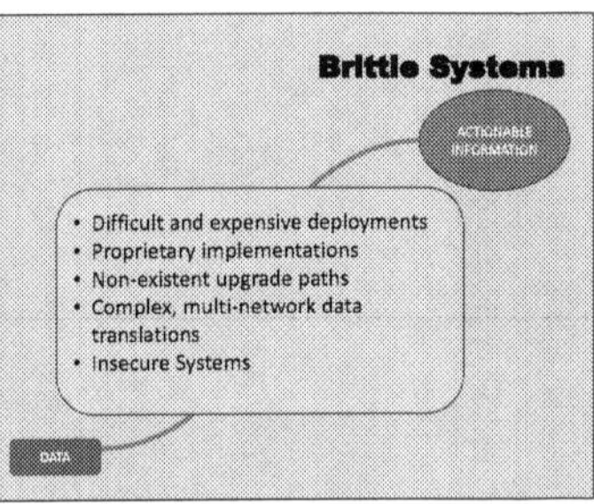

Figure 1 – Money+Equip+People=Brittle Systems

What Web Services Does

Web services is the fix for this problem. Web services defines an open, standard interface that systems can use to make some important business logic available to any other system. Some of the benefits of Web services include:

- Web services can be described, published, located and invoked over a network.
- Web services use standard, open technology available to any enterprise application.
- Web services is hassle free – you can find and use some application without downloading it, installing it and maintaining it on your computer.
- A lot of Web services are free.
- No version control to worry about. Every time you execute the Web service, you are executing the latest copy.
- Interoperability – Web service functionality is by definition interoperable with any system that can make the request.
- Language Independent – Every application can choose the programming language that best meets its requirements. The Web service request is language independent.
- Platform Independence – Web services can be supported on any platform though dedicated

automation operating systems that have to date not implemented Web services platforms as part of their product.

- Deployability – Web services are deployed over standard Internet technologies.

WS Technologies

The basic Web service technologies are HTTP and XML and SOAP.

HTTP (Hypertext Transfer Protocol) is part of the basic plumbing of the Internet. It is the low-level protocol that allows a client application like your browser to request a webpage from a web server. HTTP messages requests data or sends data in a very standard format supported by every Internet-aware application.

HTTP is a request response protocol with two basic services: GET and PUT. GET is used to request a resource (a webpage, for example) from a remote server. PUT is used to send a resource (data) to a remove server. HTTP is stateless, meaning that it is a one-time operation. There is no ongoing connection, history or memory of any previous operation.

HTTP relies on a transport protocol to move its message from the client device to the remote server. In the vast majority of cases, standard TCP (Transport Control Protocol) is used to move that message.

XML is one way that data can be structured to move between clients and servers in an open and standard way. XML encodes data in a machine and human readable way such that the data can be interpreted by any platform. Tags are the basic structure of a data item. A tag begins with a start tag like <bank balance>, is followed by the ASCII data and is terminated with a closed tag <\bank balance>. To further clarify the meanings of tags, documents called schemas are used to define the exact tags to use when sending XML documents in particular applications.

SOAP is an extension to XML that allows applications to embed remote procedure calls (RPC) in the XML documents. So now instead of just sending the dollar amount to a remote server, you can send the dollar amount and a request to convert it into Euros or Canadian dollars.

How OPC UA uses WS

For UA systems with MS Windows platforms, UA uses standard web service to encode the data and transport the message. XML is used to encode the UA client request to the server. And HTTP and SOAP are used to transport the client request to the server.

In many automation systems, this will not be the case. Though most clients will be operating in an MS Windows environment, the target server will most likely not be a MS Windows platform device. So the client device will normally use UA Binary encoding and UA TCP to encode and transport the message.

GETTING STARTED WITH OPC UA

This section is really dedicated to developers. If you develop automation products and need to use UA in your products, this section will help you figure out how to do it.

What You Need to Get Started

There are two requirements to get started with OPC UA. The first is that you have to have an Ethernet-enabled product. Without Ethernet, you can't have UA. The second is that you have to design the address space for your product.

OPC UA in an embedded automation product uses standard TCP/IP for transport. The base platform that you will need for UA is simply TCP/IP on an Ethernet-enabled processor. There is no requirement for an operating system. There is no specific requirement for processor bandwidth, flash memory or RAM.

Designing an address space will probably be the more complicated part of the project. UA uses an object-oriented address space. If your data is already object-oriented, then mapping it to UA will not be much of a problem.

But most likely it isn't.

This is a lot easier than it sounds. You need to create the object structure that represents the data for your device.

If you have a simple device, like an 8I 8O I/O device, it's going to be pretty simple. You could have something as simple as two variable type objects, each referencing a variable type node that has an array of eight bytes.

If your device is a drive with hundreds of drive parameters that you want to expose to UA, mapping that data to a UA address space is going to be a major challenge.

Option 1: Software Solution

There are a few vendors that are providing software solutions for OPC UA. Pricing, licensing and delivery for these solutions varies quite a bit. One particular vendor sells development licenses and runtime licenses. At RTA we have used a vendor that provides a binary version of UA that is licensed for a particular product.

Option 2: Hardware Solution

Board-level daughter card solutions for OPC UA. This solution will transfer data from your processor to a module that has the OPC UA address space.

Option 3: Gateway Solution

For low-volume applications, a gateway may be a better choice.

RESOURCES

There are two excellent books on OPC UA that are much more in-depth this this one. These are a good follow on to this short introduction.

Juergen Lange, Frank Iwanitz and Thomas J. Burke. OPC from Data Access to Unified Architecture– From Data Access to Unified Architecture. VDE Verlag, Berlin, Germany 2010

Wolfgang Mahnke, Stefan-Helmut Leitner and Matthias Damm. OPC Unified Architecture Textbook – The fundamentals and theory behind OPC Unified Architecture. Springer-Verlag, Berlin, Germany 2009

ABOUT THE AUTHOR

John Rinaldi, the "Dr. Phil" of Industrial Networking, is the President of Real Time Automation. Mr. Rinaldi is a leading expert in the design, development and integration of advanced networking devices in factory and building automation systems. His webinars on networking are watched tens of thousands of times each year.

Mr. Rinaldi obtained his BS in Electrical Engineering from Marquette University and a Masters in Computer Science from the University of Connecticut. At the University of Connecticut Mr. Rinaldi developed ground breaking controls for intelligent Avatars long before Avatars became popular in home gaming.

Mr. Rinaldi is experienced in all facets of Manufacturing Automation including device development (Allen-Bradley), factory floor controls (Kimberly-Clark) and enterprise systems (Procter & Gamble).

Made in the USA
San Bernardino, CA
13 June 2016